CLASSIC POST-WAR BRITISH FIRE ENGINES

ANDREW HENRY,
RONALD HENDERSON
AND JOHN TOOMEY

Captions – the captions and expert commentary for each image have been written by Ron Henderson, who is a highly respected and knowledgeable emergency services historian.

Thanks – the authors and photographers would like to thank Station Officer Bob James (retired) Glasgow Fire Service, HCB Angus historian Aidan Fisher, Colin Carter, Gary Chapman, and Andy Daley.

Front cover: This 1954 Dennis F 17/Metz TL was restored by Humberside Fire & Rescue for their museum in Hull HQ. It is now in private ownership. (Andrew Henry)

Back cover: Glasgow Fire Service placed one of the first road rescue units in the United Kingdom into service in 1972. The highly conspicuous body was built by J. C. Bennet on a 109-inch Land Rover chassis for South fire station.

First published 2022

Amberley Publishing
The Hill, Stroud
Gloucestershire, GL5 4EP

www.amberley-books.com

Copyright © Andrew Henry, Ronald Henderson and John Toomey, 2022

The right of Andrew Henry, Ronald Henderson and John Toomey to be identified as the Authors of this work has been asserted in accordance with the Copyrights, Designs and Patents Act 1988.

ISBN 978 1 3981 1150 9 (print)
ISBN 978 1 3981 1151 6 (ebook)

All rights reserved. No part of this book may be reprinted or reproduced or utilised in any form or by any electronic, mechanical or other means, now known or hereafter invented, including photocopying and recording, or in any information storage or retrieval system, without the permission in writing from the Publishers.

British Library Cataloguing in Publication Data.
A catalogue record for this book is available from the British Library.

Typesetting by SJmagic DESIGN SERVICES, India.
Printed in Great Britain.

Appointed GPSR EU Representative: Easy Access System Europe Oü, 16879218
Address: Mustamäe tee 50, 10621, Tallinn, Estonia
Contact details: gpsr.requests@easproject.com, +358 40 500 3575

Introduction

During the summer of 1973 the fire brigades of the United Kingdom were in their heyday and the envy of many other countries in the world. The Fire Services Act of 1947 had brought about the dismantling of the National Fire Service and placed the provision of fire brigades as the responsibility of county and county borough councils.

This Act had brought about the formation of approximately 150 fire brigades that covered England, Scotland, Wales and Northern Ireland. Some of these were large county fire brigades with thirty or more fire stations. However, in 1973 there were lots of small county borough fire brigades with just one fire station. All of these fire brigades were very proud institutions. Proud of their stations, fire appliances, traditions and history. Most were very protective of their fire brigade areas and extremely reluctant to call on support from a neighboring fire brigade. Many small county borough fire brigades dealt with incidents on their own rather than call on a neighboring brigade to come into their area.

Fire stations in 1973 would have fire appliances from classic British manufacturers such as Dennis, ERF, Leyland, AEC, Albion, Bedford, and Commer. Sadly, these manufacturers are all confined to the history books now. The first fire appliance that wasn't built in the United Kingdom had just been placed in service by the Glasgow Fire Service.

Today in Britain's fire stations you will find fire appliances built on Scania, Volvo, Mercedes-Benz, Iveco, MAN, Daf, and Renault chassis. Similarly, many of Britain's fire engine bodybuilders have gone out of business in recent years. There are now only a handful left in the United Kingdom.

One interesting aspect of British fire brigades in 1973 was their diversity. 150 different fire brigades bring about lots of different ideas. One of these was what colour should a fire engine be painted. Coventry Fire Brigade had famously painted their fire engines in 'Coventry Fire Engine Yellow' to aid their visibility. Several other fire brigades, such as York, Barnsley, and Wallasey, followed their example and painted their fire appliances yellow. West Riding Fire Brigade painted their fire engine fleet white. St Helens Fire Brigade painted their fleet white and red. Rochdale Fire Brigade had red fire engines with white doors.

The fire engine fleets of the United Kingdom were very varied not only in their colour schemes. Having so many different brigades meant lots of different ideas on how to design and specify a fire appliance. There were lots of interesting designs of fire appliances in service in the United Kingdom at that time.

Unfortunately, there were not many very good fire appliance photographers in the United Kingdom in 1973. Only a handful of names spring immediately to mind. In the summer of 1973 one of the best fire appliance photographers in the world visited the United Kingdom from the United States of America. John Toomey travelled from New York City to attempt to make a photographic record of the fire brigades of the United Kingdom in 1973 prior to their reorganisation the following year.

It must have been hard to organise a trip travelling from one end of the country to the other in the days before the Internet and e-mails. John travelled on public transport from London up the west coast of England through Glasgow and eventually on to Inverness. John arranged his visits to fire brigades by writing letters months in advance to the respective chief fire officer. He remembers that some fire brigades were more helpful than others. The City of Oxford Fire Brigade allowed him to stay on station. Chief Fire Officer Albert Leese of the City of Coventry Fire Brigade was really helpful.

Legendary Firemaster George Cooper of the Glasgow Fire Service was incredibly helpful, arranging for an officer to transport John around all of the Glasgow Fire Service fire stations. As a result, John made a superb photographic record of the Glasgow Fire Service, one of the most interesting in the United Kingdom. During the course of writing this book, the young enthusiastic probationary firefighter who took John around Glasgow has been identified – almost fifty years later! Bob James joined the Glasgow Fire Service in 1972. He spent twenty-four years entirely in Glasgow city stations before being medically retired as a station officer in 1996. Without his enthusiasm and excellent help in 1973, the outstanding image collection from the Glasgow area would not exist today.

Images were taken on Kodachrome colour transparency film. The unique collection remained in John's possession in the USA until 2016. The collection was not seen by many people. In 2016 I was concerned that the collection may eventually be lost forever. So I contacted John, who had been a friend of mine since the early 1970s, to ask if it could be used to make a book.

I enlisted the help of respected fire service historian Ron Henderson to assist me with photo captions. The result is a unique high-quality photographic record of the British fire brigades in their heyday in the summer of 1973.

Since that date there have been major changes to the fire services of United Kingdom. The Local Government Act of 1972 meant that many of the fire brigades of England were merged. From 1 April 1974, the approximately 124 fire brigades of England were merged into forty-seven larger fire brigades. There have been more mergers in England since 1974.

The same legislation in Wales meant that the fourteen Fire Brigade's which had been in existence from 1948 to 1974 were merged into eight brigades. In 1996 those eight brigades were merged into the present three.

Things happened at slightly different dates in Scotland. From 1948 to 1975 there were eleven fire brigades in Scotland. The Local Government (Scotland) Act of 1973 reorganised government from 1975 onwards. The eleven fire brigades became only eight. Under the terms of the Police and Fire Reform (Scotland) Act of 2012, the eight regional fire services were replaced by a single fire and rescue service for the whole of Scotland, effective from 1 April 2013.

Northern Ireland has been protected by one fire authority since 1950. The city of Belfast had its own fire brigade until 1973 when it was amalgamated into a single fire authority for Northern Ireland.

Along with the amalgamation of the fire brigades of the United Kingdom came lots of changes. As fire calls became less frequent, the fire brigades paid more attention to preventing fires and educating the public. Many fire stations were closed and it was usually claimed that there was overprovision in an area due to having separate fire brigades previously. Many appliance rooms that had six appliances in them prior to 1974 are now like a 'ghost fire station', with only two or sometimes one appliances in them. 'Special' fire appliances were taken off the run, jump-manned with another special appliance, or not even manned at all.

Recently there has been a trend to replace full-size pumping appliances with small first attendance units, in some cases with only a crew of two firefighters. In some areas two old fire stations have been demolished and replaced with one more strategically located fire station to cover the whole area.

There has become more standardisation of fire appliance types, as either two or more neighbouring brigades have made orders of the same appliances together.

The fire services of the United Kingdom today do not have anywhere near the number or variety of fire appliances as they did in 1973. These photographs represent a valuable and unique insight into that period.

Andrew Henry

The County Borough of Reading was a joint fire authority with the County of Berkshire and featured twenty fire stations. Fire appliances based on Rootes Group's Commer chassis were common features of county fire brigades. This immaculate Hampshire Car Bodies 400-gallon water tender dates from 1953 and leaves no question as to where it was based.

Another Rootes Group appliance, this time a smaller water tender based on a Karrier Gamecock chassis with bodywork by Carmichael & Sons. Pictured in the days when painting tyres was a weekly task, it is fitted with high-visibility panels and protective guards around the emergency warning systems.

The increasing obligation for attending non-fire-related incidents (special service calls) prompted many fire brigades to purchase light rescue tenders. Berkshire & Reading Fire Brigade's response was this Land Rover with engine-driven Airdrive compressor delivered in 1971.

The County Borough of Birkenhead, on the south side of the River Mersey, provided two fire stations for the protection of its properties. Displaying the borough's historical and intricate armorial bearings that were a feature of fire brigades in the pre-1974 era, this Dennis F106 water tender ladder was introduced into the fleet in 1967.

In 1974 Birkenhead Fire Brigade became a part of the newly formed Merseyside Fire Brigade. Two years before that this comprehensively equipped Ford D1617/HCB-Angus emergency tender was the last fire engine to be delivered before the merger. The illuminating 'ACCIDENT' sign and front-mounted winch are indicators of the fire services broader roles in rendering expert assistance at non-fire incidents.

After the Second World War the City of Birmingham standardised on shortened Bedford SB bus chassis for their pumping appliance fleet. Capable of operating with wheeled escape ladders, Prestage did the fire engineering and Wilsden did the coachwork. There was a total of twenty-seven in the fleet; this one dating from 1952 was the second one to be commissioned.

After the Bedfords, appliances from Dennis Bros formed the next generation of the cities pumping appliances. Commencing in 1965, eighteen Rolls-Royce-powered F36 dual-purpose appliances were ordered; this one from Aston fire station is one of the last of the batch.

Birmingham's next series of Dennis pumping appliances were built to the same pattern as the earlier F36 machines but were based on Dennis F45 chassis. Ordered in two batches of five, this one based at Handsworth was the last of the first batch from 1970.

Seven Dennis dual-purpose appliances, this time based on F48 Series chassis, were to become the last of the type delivered to the city before the brigade was incorporated into the West Midlands Fire Brigade in April 1974. One of a batch of seven delivered in 1972, this was another one that was based at Handsforth.

Birmingham's pre-war and war-time turntable ladders were initially replaced by a single AEC Regal with Merryweather equipment followed in 1956 by this smart Bedford appliance fitted with a 100-foot Magirus turntable ladder and coachwork by Wilsden. This one operated from the city's Aston fire station and fortunately still survives.

In 1961 a further two 100-foot Magirus ladders, this time with hydraulic operating systems mounted on Bedford TK chassis, were delivered to Birmingham. Right up until April 1974 the city fire brigade also operated the ambulance service and one of the service's Bedford J1 ambulances can be seen in the background.

Birmingham's turntable ladders were supplemented by two hydraulic platforms, the first one being this Jennings-cabbed ERF with an 85-foot Simon elevating platform delivered in 1969. The second one, delivered in 1973, was Dennis-based.

Another classic example of Britain's coachbuilding expertise. Dating from 1956, this Wilsden-bodied Bedford breathing apparatus tender started life as a rescue-breakdown vehicle complete with Harvey Frost crane mounted on a platform at the rear. It is shown here after the crane was transferred onto a new Dennis chassis.

In the absence of a salvage corps in Britain's second biggest city Birmingham operated this unique Homalloy-bodied salvage tender based on a 1963 Bedford TK chassis. It was based it at the city's headquarters fire station.

One of the characteristics of the early post-war fire brigades was the variety and individual styling of appliances, many of them built to unique designs. Birmingham's Commer 1½-ton control unit delivered in 1960 was built by renowned ambulance builder Herbert Lomas Ltd, of Wilmslow.

Custom-built breakdown lorries were rare features of Britain's fire service but Birmingham was one of the exceptions. The 8-ton Harvey Frost crane previously mounted on the Bedford breathing apparatus tender was transferred to this shortened Dennis F37 chassis.

The citizens of the noted holiday resort of Blackpool were protected by three fire stations with a mixed fleet of fire engines. The workhorses were the first turn out appliances equipped with a wide array of firefighting and ladder equipment. One example was this Hampshire Car Bodies (HCB)-Bedford TK pump escape delivered in 1961.

AEC fire engines were always smart looking, as proven here with this AEC TGM/HCB-Angus pump escape fitted with a Merryweather 50-foot wheeled escape ladder and rear pump with controls mounted on the sides. The mounting of the hose reels on the roof allowed for more locker space for small items of equipment.

Blackpool's last fire engines before the brigade's incorporation into Lancashire County Fire Brigade in 1974 were based on ERF chassis. This one, although built to the design of a water tender and fitted with a pump, also combined the functions of an emergency tender. Bodied by HCB-Angus, this 1972 example was the first of the town's three ERFs and operated from Bispham fire station.

From 1966 Merryweather & Sons replaced their first range of AEC Mercury 100-foot turntable ladders with the new TGM Mercury range incorporating an Ergomatic cab. Blackpool Fire Brigade was the first English brigade to commission one, taking delivery of this one in 1968.

In 1972, in order to supplement the town's two turntable ladders Blackpool Fire Brigade took delivery of this ERF 84RS appliance fitted with Simon Engineering 70-foot hydraulic platform. It was unusual to see such an appliance equipped with a 'first-aid' hose reel.

Bolton Fire Brigade commissioned the whole range of Merryweather's Regent and Marquis fire engines. This comprehensively equipped AEC clone, delivered in 1959, is actually a Marquis Series III based on a Maudslay Merlin 2 chassis. As with most of the fleet it became part of Greater Manchester Fire Brigade in 1974.

Although similar to the previous photo, this one dating from 1963 is a Series V Merryweather Marquis, this time on an AEC Mercury chassis. The longer 13-foot-6-inch wheelbase is a good recognition point. The long roller shutter locker was designed to house a portable pump and, on the roof, a Total foam-making branch is carried.

The AEC TGM appliances from Merryweather were the last of the range. Delivered in 1970 and equipped with a Merryweather 50-foot steel-wheeled escape ladder, it was one of two in service in Bolton. The other one, delivered two years later, was written off shortly after delivery after being involved in a collision. The front-mounted orange blinker light is an unusual accessory.

This 1973 Dennis F108 Series water tender ladder was the borough's last new fire engine before the brigade was incorporated into the newly formed Greater Manchester fleet. By this time the traditional wheeled escape ladder had been replaced by a 45-foot light alloy 'Lacon' ladder.

In 1956 Merryweather & Sons introduced a new range of 100-foot hydraulic turntable ladders that quickly gained popularity with fire authorities both at home and overseas. Bolton received this one in 1966. The large 'FIRE BRIGADE' legend was a later feature of the brigade's appliances.

ERF introduced their first fire engine chassis in 1966, one of the ranges being designed initially to carry a Simon Snorkel hydraulic platform ranging in height from 65 feet to 85 feet. They were big, heavy machines as seen here in Bolton's 85-foot example, delivered in 1971.

Emergency Tenders were special appliances equipped with a wide range of rescue equipment, including emergency generator and lighting equipment and additional sets of breathing apparatus. This Bedford TK with coachwork by Dennis Bros, under the Dennis 'M' marque, was delivered to Bolton in 1966.

Most of Bootle's post-war pumping appliances were based on Dennis chassis. This F35 Series appliance dating from 1967 features low mounted two-tone horns and bell and is carrying a Merryweather 50-foot wheeled escape ladder.

Also fitted with low mounted emergency warning systems, this 1972 Dennis water tender was the last new fire engine bought by Bootle Fire Brigade before the brigade was incorporated into Merseyside Fire Brigade in 1974.

Bootle was on the north side of the River Mersey and was covered by two fire stations. As with most county borough fire authorities it featured an eclectic array of fire engines. An unusual characteristic was the brigade's title represented in scroll lettering as seen here on the brigade's 1957 Bedford S Type Wilsden-bodied vehicle with 100-foot Magirus ladder and built-in pump.

Swedish-made 'Orbitor' hydraulic platforms were a rare feature on UK fire appliances compared to those produced by Simon Engineering. This smart Dennis F119 machine, supplied by Carmichael & Sons of Worcester and delivered in 1970, featured 72-foot booms.

Specialised emergency tenders were features of many fire brigades and Bootle's choice was this Bedford TKEL supplied by HCB-Angus Ltd in 1972. The red and white chequered cylinder on the roof signifies the vehicle's dual function as a fire ground control point when operating at large incidents.

Burnley was one of the smaller county borough fire authorities with just one fire station. Among its small fire engine inventory was this classic 1963 Leyland Super Six water tender supplied by John Morris & Sons of Salford. Equally unusual was the pair of large trumpet horns on the roof, which were part of the warning bells amplifier system.

In 1964 the Leyland water tender was supplemented by a similar appliance but modified to carry a 50-foot Simon hydraulic platform and a limited range of equipment normally carried on an emergency tender. These machines remained rare models in the UK with the only other being a pump escape operated in Dudley, West Midlands.

This 1961 Land Rover-Carmichael Redwing light pumping unit was overlooked by Burnley's licencing department by not incorporating a single number '1' in the vehicle's registration. Ideal for incidents in parts of Lancashire County that the brigade covered, it featured a built-in 250 gallons per minute pump and front-mounted winch.

The town of Bury in Lancashire had one fire station with a mixed fleet of appliances. This 1973 ERF/HCB-Angus appliance with 50-foot hydraulic platform was the last order for the borough before it was incorporated into the newly formed Greater Manchester county in 1974.

Bury took delivery of this Dennis F107 appliance with 100-foot hydraulic Magirus ladder in 1965. Nearby Oldham Fire Brigade operated a similar model. Several units in the fleet featured registrations incorporation the numbers '999'.

Coventry Fire Brigade was the instigator of one of the most profound changes in post-war history when it introduced yellow fire appliances that experiments showed would be more visible at night, especially under sodium street lighting conditions. This 1966 Dennis F36 and two sisters ordered at the same time were the first.

The pioneering trio of Coventry yellow fire appliances were dual-purpose appliances, originally equipped to operate with a wheeled escape ladder or wooden 'Ajax' extension ladder. They were equipped with dual-purpose hose reels for either fire extinguishing or powering compressed air tools.

Following the receipt of the prototype yellow Dennis appliances most but not all of the existing vehicles were repainted in the pioneering yet garish 'Coventry Yellow' livery. There were four of these Rolls-Royce-powered Dennis F12 pump escapes in the fleet, all of which acquired the new livery.

Another of Coventry's Dennis F12 appliances after a repaint job. Even the Merryweather 50-foot steel-wheeled escape ladder has not escaped the repaint process. This one, dating from 1956, operated from Canley fire station.

Another that underwent a full makeover, this classic Dennis F21 was one of three turntable ladders in the post-war fleet. The chain drive mechanism on the ladder turret was a characteristic of the early post-war German-made Metz ladders.

Another makeover and what a resplendent one of Coventry's Dennis F12 emergency tender. This one had the additional functions of carrying additional breathing apparatus and fire ground control facilities hence the red and white chequered band at the roof line. This appliance is one that still survives.

In 1972 the city's allegiance to Dennis was changed when a pair of Rolls-Royce-powered Bedford vehicles with HCB-Angus bodywork were commissioned, followed by a third the following year. These were the last appliances bought before the fleet was incorporated into West Midlands Fire Brigade in 1974

In 1964 the city received this Dennis F37 appliance with German-made Magirus 100-foot ladder. Only a few of this pattern saw service with UK fire brigades but it looks resplendent with its red and silver livery. This one managed to avoid the painter and his yellow tin of paint.

Coventry had three fire stations and featured a very progressive fire brigade with a mixed fleet of appliances. A total departure from the usual larger city fire appliances was this small but comprehensively equipped short-wheelbase Land Rover/Carmichael Redwing FT1 light pump delivered in 1963.

Not all fire appliances were constructed to elaborate designs as is evident by Coventry's 1963 Morris FE hose laying lorry. This appliance, capable of paying out lengths of hose while underway, also escaped the artisan with his yellow paint.

The County Borough of Darlington Fire Brigade was another of the smaller but proud and well-equipped fire brigades whose fire and ambulance fleet featured '999' registrations. Operating from one fire station, this Dennis F106 water tender, delivered in 1965, was the workhorse of the fleet. (Andrew Henry)

Darlington Fire Brigade was one of the pioneers in setting up a scheme for rendering swift action to victims involved in road collisions, commissioning this Dennis Maxim breakdown lorry fitted with a Dial-Holmes twin boom hydraulic winch. In 1974 the Darlington fleet became part of Durham County Fire Brigade. (Andrew Henry)

Of the ten Leyland Firemaster appliances built to the order of UK customers, two of them were fitted with 100-foot Magirus ladders. Unlike the Wolverhampton example, Darlington's did not have a front-mounted pump. Dating from 1959, it became part of the Durham County Fire Brigade fleet in 1974 and is another classic appliance that still survives. (Andrew Henry)

Durham County Fire Brigade was another authority committed to providing a comprehensive rescue facility on the nation's increasingly congested road system. In 1960 they received this coach-built Hampshire Car Bodies 4x4 Bedford RL breakdown vehicle with Harvey Frost crane, seen here at the county's Framwellgate Moor headquarters. (Andrew Henry)

Glasgow was unique as it was the sole Scottish fire authority that retained its pre-war identity. It had one of the most varied fleet of any fire brigade and its appliances are consequently featured much in this book. Upon the brigade being subsumed into Strathclyde Fire Brigade in 1975 this 1956 Dennis F12 pump escape had given almost twenty years of hard work in the city.

There were some smaller F8 type Dennis pumping appliances in the Glasgow fleet and then in 1958 a pair of F24 Series Dennis pump escapes were delivered. They were the least photographed of Glasgow's fire appliances but luckily this one from Castlemilk fire station has been captured perfectly.

Leyland Motor Company's novel 'Firemaster' appliance featured the engine mounted between the chassis frames, just ahead of the rear wheels and the main pump situated at the front enclosed by two doors. Only ten were supplied to UK customers. A futuristic appliance for the time, this one dating from 1959 was the first of two for Glasgow.

Glasgow received their second Firemaster in 1960 and although carrying the same equipment it had a redesigned cab of a much more rounded appearance. The first Firemaster operated from Partick and is now on display in the cities Riverside Museum. The second operated from North West fire station in the Kelvingrove district of the city.

There were several ACV Group appliances in the Glasgow fleet including a pair of AEC clones that were in fact based on Maudslay Merlin 2 chassis with AEC's AV470 diesel engine. Bodywork and fire engineering was designed by Merryweather & Sons, which marketed the type under the Marquis marque.

Between 1962 and 1963 Glasgow received nine of these AEC Mercury's that were unique to Glasgow. Some of them carried wheeled escape ladders but this one is equipped with 45-foot, 30-foot and short extension ladders. The extended front bumper provided a convenient step to facilitate cleaning of the ample windscreens.

Among Glasgow Fire Service's many accolades was the distinction of being the first UK fire authority to commission appliances built on a foreign-made chassis. Four of these splendid German air-cooled Magirus-Deutz pump escapes and two Magirus turntable ladders entered service in the city in 1968. Edinburgh firm Scottish Motor Transport built the bodies on the pump escapes.

Another unique class of appliance, developed in conjunction with the Glasgow Fire Service, the 'scoosher' was designed initially to overcome weight and width restrictions on the newly developed Anderston Centre. It featured an elevating boom that could project water into the upper reaches of a building. This is the Mark I version built by Bennets on a Dennis D chassis and dates from 1969.

In 1970 a further batch of five scooshers was ordered to complement the three previous examples. These latter ones were constructed on Dennis F46 chassis and improvements comprised of the turret of the elevating booms being mounted at the rear and the facility of the 45-foot ladder able to be operated without having to be removed from its attachment on the boom.

Glasgow Fire Service bought a pair of these Dennis F14 appliances with 100-foot Metz ladders. The first one delivered in 1954 is seen outside the South fire station before being relocated to Partick. The second one, delivered in 1957, was tragically destroyed in the Cheapside fire in 1960.

Following on from the original six Magirus-Deutz appliances of 1968 one final German vehicle was bought in 1970. Also fitted with a 100-foot Magirus ladder this final appliance was fitted with the innovative addition of a cage at the head of the ladder, the first UK example and third in the world with this feature.

A spectacular machine with an interesting history. This 1960 AEC Mercury pump-hydraulic platform appliance originally carried a Merryweather 100-foot turntable ladder that was destroyed when the appliance turned over while the ladder was extended. The chassis was salvaged and fitted with a new Bennet body and Simon Engineering hydraulic platform.

Even more spectacular and one of the smartest of the post-war fire engines. A truly multi-purpose appliance built by Bennets on a 1969 AEC TGM Mercury chassis. It featured a 1,000 gallons-per-minute pump with an in-built foam system and a 70-foot Simon Engineering hydraulic platform. It operated from the city's North West fire station.

The risks inherent in Glasgow ensured the fire service fleet had a comprehensive array of special appliances such as this impressive Carmichael-bodied AEC Mercury emergency tender from the South fire station. It was one of a pair ordered in 1963, the second being later converted into a breathing apparatus tender.

The construction of an urban motorway in Glasgow prompted the introduction of a pair of Land Rover Road Rescue Units. The first of the pair delivered in 1971 is seen here. As well as a comprehensive inventory of rescue and cutting tools, they also featured a hydraulic winch, electric generator and Halogen floodlights on a telescopic mast.

Another of Glasgow's classic fire appliances and again unique to that city, this is one of a pair of AEC Mercury's supplied by Pyrene & Sons of London. It was capable of producing 6,400 gallons per minute of foam from its 1,000 gallons per minute pump and in open side lockers there were twelve 20 lb dry powder extinguishers.

Another foam tender but a total contrast to the big AECs. High-expansion foam was a new concept in the 1960s and Glasgow was one of the first cities to appreciate its advantages. This Morris FG/Bennet unit, delivered in 1968, carried the foam equipment and towed the large generator needed to produce the finished foam.

Salvage Corps were vestiges of the old insurance fire brigades and were funded by the insurance companies to attend fires and salvage any property affected from fire, thus reducing the fire damage costs. Glasgow operated one of the UK's three such corps with this uniquely designed 1964 Bedford TK being one of the mainstays of the small fleet during the 1960s.

Another of the Glasgow Salvage Corps vehicles, this short-wheelbase Bedford TK dates from 1968. These early Bedfords were replaced by three new Fulton & Wylie-built Bedford TKs that were short lived as in 1984 the insurance companies disbanded their salvage functions in Glasgow, Liverpool and London marking the end of such a noble and historic organisation.

The fire service fleet in Grimsby was almost wholly based on Dennis appliances and seen here is the town's fourth new post-war appliance, an F24 series appliance delivered in 1958. Of dual-purpose configuration, it could also carry a wheeled escape ladder. (Andrew Henry)

In the late 1960s a novel concept was to affix a 50-foot hydraulic platform to a conventional pumping appliance, combining the functions of rescues from buildings and firefighting capabilities. Grimsby commissioned this Rolls-Royce-powered F36 series pump-hydraulic platform vehicle in 1967. (Andrew Henry)

The Borough of Grimsby's last new appliance was this F49 series Dennis pump escape, delivered in 1973 – just one year before the brigade and its fleet were incorporated into the newly established Humberside Fire Brigade. (Andrew Henry)

Looking immaculate in the morning sunshine, Grimsby took delivery in 1954 of this Dennis F17 appliance fitted with a 100-foot German Metz turntable ladder and originally a self-contained pump mounted aft of the open-backed cab. Restored by the brigade for posterity, it still survives in the hands of a private owner. (Andrew Henry)

Dating from 1961 this economical adaptation of a forward-control Commer 1½-ton van equipped it with the dual function of operating as a rescue tender or a fire-ground communications unit. (Andrew Henry)

Bedford's TK vehicles formed the basis of many fire appliances in the 1970s especially with the county fire authorities. More commonly seen with unpainted aluminium stucco bodies, this rare Rolls-Royce-powered example, dating from 1971, was one of three that served with Holland County Fire Brigade, Lincolnshire. (Andrew Henry)

There were five of these Austin FGK appliances equipped by Carmichael & Sons as hose reel tenders for use at the counties more rural fire stations. Fitted with an in-built hose reel pump, a trailer pump also formed part of the unit. They were unique to Holland County and one of them still survives. (Andrew Henry)

West Riding of Yorkshire Fire Brigade chose the colour white for their fire appliance fleet on account of the greater visibility in darkness and low-light conditions. Holland County followed suit with this 1971 Land Rover/Carmichael rescue tender. (Andrew Henry)

The port town of Boston was the most populous in Holland County and was where this 1972 HCB-Angus bodied ERF with Simon 70-foot Snorkel hydraulic platform appliance was based. This one was also powered by a Rolls-Royce engine. (Andrew Henry)

Many British fire brigades started the post-war years with a plethora of wartime standard fire appliances. This half-cab Leyland TD7 bus chassis was one of the wartime adaptations fitted with a Merryweather 100-foot ladder. One of the last of the type in service, part of the driver's inventory included leather helmet, gauntlets and goggles. This vehicle is still extant, located in Holland. (Andrew Henry)

In 1949 ten AEC Regal vehicles with 100-foot Merryweather ladders were commissioned by the Home Office for allocation to brigades with the greatest need for replacing obsolete pre-war equipment. One of the ten was allocated to the City and County of Kingston upon Hull Fire Brigade, as is displayed on the mahogany name plate on the side of the ladder. This machine still survives. (Andrew Henry)

Lanarkshire was one of the counties where fire engines were seldom photographed, so this photo of the county's sole Rolls-Royce-powered Dennis F8, a smaller version of the F12, is particularly worthy of inclusion. Delivered in 1953, it was assigned to the part-time fire station at Bishopbriggs.

This small-sized Bedford water tender, built by Hampshire Car Bodies, was a common feature of some rural county fire brigades. Dating from 1957, this example allocated to Shotts has the unusual feature of a separate door for the driver. Access was usually through the same sliding doors as the rest of the crew.

Apart from a few Bedfords, the Lanarkshire fleet of pumping appliances was predominantly supplied by Dennis Bros, typified by this 1969 F45 model. Fitted with a 50-foot Merryweather steel-wheeled escape ladder, it was one of a pair based at Cambuslang fire station.

Following the delivery of this Dennis F48 in 1973 there were only two more appliances of identical pattern delivered before the county fire brigade was incorporated into Strathclyde. By this time the traditional wheeled escape ladders were being replaced by 45-foot light alloy ladders.

In Scotland there were two of these AEC TGM Ergomatic cabbed appliances with 100-foot Merryweather ladders. Fife Fire Brigade took the first UK example in 1967 and Lanarkshire took the eleventh, allocating it to Cambuslang station.

Lanarkshire Fire Brigade's headquarters was at Hamilton where the associated fire station operated a turntable ladder and this king-sized 85-foot Simon Snorkel mounted on an ERF 84RS chassis. It will be noted that a feature of the Jennings cab comprises of a pair of front assemblies mounted back-to-back.

The Simonitor appliance consisted of a hydraulically operated set of booms with a monitor at the head used for projecting large quantities of water into the upper floors of buildings. Only ten of this design saw service in the UK, with Lancashire County Fire Brigade buying two Ford D1616/HCB-Angus models in the early 1970s for their Chadderton and Leigh stations.

Liverpool Fire Brigade was the first in England to commission one of the new Simon Snorkel appliances, receiving this unusual cabbed Dennis F117 in 1963. Fitted with 65-foot booms and a midships-mounted pump, it was joined by a second one four years later. The particular design of the Dennis cabs remained unique in the UK.

The AEC Regent III Merryweather dual-purpose appliance formed the mainstay of London Fire Brigades fleet during the 1950s and 1960s with a total of fifty-two being bought. Fitted with 1,000 gallons per minute pumps and designated as dual-purpose appliances, this one operated from Tottenham fire station.

Next to the AECs Dennis fire appliances predominated in London, most of them built to a design unique to London. Typical of them was this Dennis F106 dual-purpose appliance equipped with a midships pump and 50-foot Bayley-wheeled escape ladder. The J21 plate on the door identifies its base as Edmonton.

The formation of Greater London Council in 1965 and the amalgamation of surrounding fire brigades saw London acquire many non-standard fire appliances. This 1962 Bedford TK with HCB Angus bodywork was one such example. Inherited from Middlesex, it was equipped to carry a wheeled escape ladder when necessary and was the only appliance built to this design.

This 1961 Dennis F26 dual-purpose appliance was one of several inherited from Essex County Fire Brigade during the 1965 merger. The merger resulted in silver fire appliances operating in the city for the first time as this Barbican-based example demonstrates.

London Fire Brigade and West Riding of Yorkshire Fire Brigade specified retracted front axles and four door cabs on their AEC turntable ladders. This Paddington appliance, one of six delivered in 1966, was one of eighteen of the type supplied to London by Merryweather & Sons.

This emergency tender was one of three of this design in the London fleet, with another at Liverpool, that were the biggest such appliances in the UK. This one based on a Dennis CV31 chassis operated from Paddington fire station.

A former West Ham Fire Brigade 1959 Maudslay Merlin II/Merryweather Marquis emergency tender. After incorporation into the Greater London fleet in 1965 the brigade's emergency tenders were adorned with 'RESCUE' titles to promote the increasing non-fire-related responsibilities.

Another odd one in the London fleet. An unusual Merryweather Marquis foam tender fitted with a non-standard cab and based on a Maudslay Merlin 2 chassis. It was another inherited from West Ham Fire Brigade and is seen when assigned to Edmonton fire station.

Purpose-built breakdown lorries were not common features of UK fire brigades but London always featured one in their fleet. This shortened Dennis F107 chassis was fitted with the crane from an earlier breakdown lorry.

This rear view shows the original Herbert Morris 5-ton crane, transferred from the brigade's pre-war Dennis chassis. Displaying B21 plates representing Clapham fire station, it still survives, although the original crane has since been replaced by a Harvey Frost one.

Another appliance not typical to the London fleet. One of two such Sun Engineering-built 'parrot-nosed' Dodge Kew foam tenders delivered new to Essex County Fire Brigade in 1958 and inherited by London in 1965. The L21 plate signifies it being based at East Ham Divisional Headquarters.

The illustrious London Salvage Corps celebrated its centenary in 1966, one year before this Wood & Lambert-built Ford Thames Trader salvage tender was delivered. The three salvage corps, funded by insurance companies, were a welcome adjunct to the firefighting services in Glasgow, Liverpool and London but were all disbanded in 1984.

Leyland Motors re-entered the fire engine market in 1959 with the introduction of their 'Firemaster' chassis, which featured the engine mounted amidships between the chassis frames and the pump mounted behind doors at the front. Manchester Fire Brigade received the prototype in 1959, the first of three in their fleet.

Manchester's second Firemaster appliance complete with wheeled escape ladder, seen when operating from Philips Park fire station. The 30-foot 'Ajax' extension ladder housed at an angle was a Manchester characteristic. For several years a matched pair of these appliances was based at the London Road headquarters fire station.

Between 1966 and 1972 Manchester received fourteen of these Albion Chieftains equipped as either pump escapes or major pumps. Four different bodybuilders were involved with this penultimate example from Philips Park station, featuring bodywork by Carmichael & Sons.

This was the second of two classic AEC Mercury's with Merryweather's 100-foot turntable ladder bought by Manchester Fire Brigade. Delivered in 1961, it was based at the London Road headquarters fire station. Both of Manchester's AEC turntable ladders had the unusual feature of a shorter cab with single hinged cab doors as opposed to the standard sliding doors.

Another unique and undoubted classic was Manchester's first hydraulic platform with its garish yellow but highly visible 65-foot Simon Snorkel booms. Delivered in 1963 and assigned to London Road fire station, Cockers of Southport did the coachwork.

Another garishly coloured Simon Snorkel for Manchester but mounted to an ERF 84PS chassis. Delivered in 1972, this HCB-Angus-bodied example had 85-foot booms and unusually featured a two-man cab as opposed to the more usual and larger four-door cab. This one operated from Moss Side fire station.

An impressive example of a classic large style emergency tender and more classically based on a Leyland Firemaster chassis. Of the ten UK Firemasters, Manchester operated four but this emergency tender was the only one built to this design. The front locker contained a Capstan winch in place of a pump.

Another appliance unique to Manchester this underfloor-engined Albion Claymore was one of a pair built by Cocker and equipped as a salvage tender to provide facilities for reducing fire and water damage to property during and after a fire. This one was based at the Manchester suburb of Blackley.

Coach-built foam tenders with built in foam-making facilities were comparatively rare features in local authority fire brigades and fortunately one of the least used fire appliances but a necessary adjunct in the event of large-scale petroleum and oil fires. Manchester commissioned this Albion example with Pyrene equipment in 1964.

The town of Northampton, with its single fire station at The Mounts, was another locality whose fire appliances were seldom photographed. However, as was common at the time, the appliances were kept in pristine condition. This 1957 Dennis F12 with rear-mounted 1,000 gallons per minute pump was the first turn out appliance for many years.

Standard HCB-Angus Bedford TK water tenders were more often seen in service with county fire authorities owing to their 400 gallons water tanks. Nevertheless, it would have been a useful adjunct for rural fires. The multitude of warning systems on the roof of this 1962 Bedford TK includes a set of four Bosch electric horns.

Before being incorporated into the Northamptonshire County fleet in 1974 the city fire brigade ended its independence with a pair of Dennis F48 water tenders commissioned in 1972. The immaculate condition of this second one of the pair suggests it has just been delivered.

For a short time, Northampton Fire Brigade followed Coventry's pioneering experiments and painted some of its appliances in yellow livery. For attending road collisions and other special service calls where rescue equipment was required the brigade adopted this 1971 Bedford CF van as their rescue tender.

Another of the classic AEC Mercury/Merryweather turntable ladders but of particular interest is that this was the last of the line before Merryweather adopted the successive AEC TGM ergomatic cab for its next range of appliances. Delivered in 1967 and at one time furnished in an all-over yellow livery, this one still survives.

As far north as one can go, on the Scottish islands the Northern Area Fire Brigade of Scotland governed one full-time fire station and fifty-eight part-time and volunteer stations. This comprehensively equipped 1961 Commer with HCB bodywork operated from the brigades only full-time station at Inverness.

One of the larger appliances in the Northern Area fleet, the delivery of this Dennis F108 water tender in 1970 was undertaken to protect the citizens and properties of the region's biggest city, Inverness. It was the only one of its type in the brigade.

Northern Area operated six of these small Bedford A3 hose reel tenders delivered in the mid-1950s for the more isolated areas such as Lerwick, Stornoway and Portree, etc. Judging by the broom on the roof this one pictured at Inverness has been relegated to secondary duties such as fire salvage work.

Carmichael & Son's six-wheel Commando conversion of the Range Rover was widely adopted by public, private and government fire services for use especially as a light pumping unit or rescue tender. This one with front-mounted pump was assigned to Northern Area's Inverness fire station in 1972.

Northern Area was one of the first fire brigades to commission an ERF appliance with 85-foot Simon Snorkel booms when they received this Fulton & Wylie-bodied example in 1967. The *MacIntyre* tartan stripe at the front was a unique addition applied at the behest of the brigade's Firemaster, Eric Macintyre M.I.Fire E., QFSM.

Northern Area Fire Brigade did a good job of refurbishing this former Auxiliary Fire Service green goddess. Acquired from one of the government sales of surplus civil defence equipment, the drums of foam on the roof suggest that it was acting in the role of a foam tender when pictured on the forecourt at Inverness.

Oldham Fire Brigade with its two fire stations was another user of the renowned 1950s Dennis product, the Rolls-Royce-powered F12 machine. One of a pair, this one looking rather the worse for wear had been in service for almost twenty years when pictured at the headquarters station.

This one was an odd ball. Although built in 1963 by HCB, it was built to a design unique to Oldham, especially concerning the lockers, which normally had roller shutter covers. It was a dual-purpose appliance capable of operating with a wheeled escape ladder. Protruding from the front grille is a Francis long rolling American type siren.

Following the delivery of the Bedford TK appliance Oldham reverted to Dennis products, taking delivery of this F36 series pump escape in 1967. Around this time huge reverse FIRE letters were affixed to the front ensuring maximum visibility in motorist's rear-view mirrors.

Forsaking Dennis again the next pumping appliance in Oldham's fleet was this standard Carmichael-bodied 400-gallon water tender/ladder mounted on a Ford D600 chassis. This one and most of Oldham's post-war fleet passed into Greater Manchester in April 1974.

The last two appliances ordered by Oldham Fire Brigade were both Dennis's, the penultimate one being this F48 water tender ladder delivered in 1972. By this time wheeled escape ladders were rapidly being replaced by 45-foot light alloy ladders and here the ladder is complemented by a 35-foot light alloy extension ladder.

Turntable ladders mounted to Dennis F107 chassis were uncommon features in the UK but Oldham and nearby Bury both operated this type complete with Magirus 100-foot ladders. Oldham's was received in 1964.

In 1971 Oldham supplemented their high-rise turntable ladder with this more versatile 70-foot Simon Snorkel hydraulic platform mounted on a Jennings-bodied ERF chassis. The twin hose reels were an unusual feature of this appliance.

Oldham had a pair of these Miles-bodied Bedford TJ appliances, the first a water tender and second this emergency tender with clerestory roof extension to increase the interior headroom. An optional feature of Miles appliances was rubber mudguards that helped to reduce the workload of the council's coach builders and painters.

A classic machine from the 1950s, this 1,000 gallons-per-minute AEC Regent III/Merryweather major pump was the first new post-war delivery to the Oxford City Fire Brigade and was originally fitted with a Merryweather 50-foot wheeled escape ladder when delivered in 1951. It bears the registration 'SFC1' that would later be classed as a much desirable registration number.

More often seen in operation with county fire authorities Oxford bought this Carmichael-bodied Karrier Gamecock water tender ladder in 1961. The assortment of ladders includes a Merryweather 45-foot light alloy ladder and an unusual aperture for the business end of the hose reel.

After purchasing pumping appliances based on AEC, Bedford and Karrier chassis the city's next new pumping appliance was Dennis-based, with the delivery in 1965 of this AEC AV470-engined F106 model.

In 1972 Oxford's penultimate Dennis appliance was based on an F108 chassis. Following the delivery of the final city appliance, a smaller Dennis D model, two years later the city service was amalgamated with the county of Oxfordshire.

Oxford City was another of the fire brigades that took up the option of acquiring redundant Auxiliary Fire Service vehicles. This 1954, 4 x 2 former green goddess emergency pump was given an unusual livery at a time when Britain's fire brigades were experimenting with more conspicuous liveries.

Dating from 1962 Oxford City was one of the many British fire brigades that adopted Merryweather's successful 100-foot hydraulic-operated turntable ladder. This AEC Mercury appliance is fitted with a built-in pump, but unusually it also has a hose reel, mounted below the pump locker.

Another of Oxford's dual-coloured appliances, a 70-foot Simon hydraulic platform unit mounted to an ERF chassis, a combination that was as equally successful as Merryweather's AEC ladder combination. This unit, also fitted with a hose reel, dates from 1970.

Custom-built fire appliances using Ford's Thames Trader chassis were decidedly rare in the UK but widely adopted in parts of mainland Europe. This 1963 Hampshire Car Bodies emergency tender was unique to Oxford and definitely a 'classic'.

Rochdale consisted of one fire station with a varied fleet of appliance makes. The brigade's two earlier Dennis appliances were increased by the delivery of this Dennis F36 water tender/ladder in 1963. As is evident the brigade followed Teesside Fire Brigade's lead and adorned the fronts of their appliances with white paint.

Dennis's early post-war turntable ladders were built on similar chassis to the F12 pumping appliances but with longer wheelbases. The dark green German-made Metz ladder is mounted on an F21 chassis and at 125 feet in length the ladder was one of the tallest in the country and still survives at Manchester's Fireground Museum.

Ten years passed between the delivery of the 1963 Dennis water tender/ladder and the next pumping appliance, the first of two ERF 84PF/HCB-Angus water tender ladders.

The pair of ERF water tenders were joined by this combination emergency-salvage tender and control unit, hence the chequered roof panel. Shortly after this appliance was delivered the town's chief fire officer was appointed to a similar position but in a larger brigade at Sunderland and promptly ordered an identical appliance for that town.

Salford Fire Brigade had one fire station but as with most county borough brigades it featured an interesting array of fire appliances. This two-tone red Dennis F12 delivered in 1954 was the only one built to this particular design.

The sign on the front of the roof leaves it in no doubt as to who operates this appliance. Delivered in 1970, it is a Rolls-Royce-powered Dennis F46 water tender ladder equipped with a Merryweather 45-foot 'gut buster' light alloy ladder.

In 1968 Salford commissioned this HCB-Angus-bodied ERF84RS 65-foot Simon hydraulic platform, the specifications of which included the unusual features of a hose reel and small water tank.

In 1974 Salford was incorporated into the newly formed Greater Manchester Metropolitan County. The city's penultimate fire appliance was this Dennis F48 combination emergency/salvage tender delivered in 1972.

Solihull had a relatively short-lived fire brigade, formed in 1964 from parts of Warwickshire. This impeccable 1956-built Carmichael-bodied Commer water tender ladder was inherited from Warwickshire.

This Rolls-Royce-powered Dennis F38 water tender delivered in 1969 was the second of the type in the Solihull fleet.

Delivered in 1965 just after the brigade was formed, Solihull chose Bedford's TK chassis for their emergency tender contracting HCB-Angus to construct the body.

It is pleasing to note that Solihull Fire Brigade skimped on the unpainted aluminium bodywork notion, opting instead for traditional fire engine red. This 85-foot Simon Snorkel has bodywork by HCB-Angus and is complete with hose reels and a 80-gallon water tank.

Following West Riding of Yorkshire Fire Brigade's adoption of an all-over white livery St Helens followed suit but added red highlights. This 1962 Dennis F28 pump escape is shown after its refurbishment. The traditional wheeled escape ladder has been replaced by a 45-foot light alloy ladder and the brigade's characteristic king-sized FIRE sign has been added.

This Leyland Beaver water tender represents the epitome of early 1970s fire engine developments. One of three at St Helens, this Carmichael & Sons-built appliance was the first of the three and among several innovations is the heavy-duty crash bar on the front.

Another of St Helen's uniquely liveried appliances. This AEC Mercury with Merryweather 100-foot ladder is another that was all red when delivered to the brigade in 1959. Unlike the Leyland water tenders this one retained its white livery following the incorporation of the brigade into Merseyside in 1974.

St Helens high-visibility white and red livery seems to work well on this 1967 Albion Chieftain/Carmichael rescue tender, another one that was repainted from its original red livery. The machine has a bumper-mounted loud hailer and follows Lancashire's trend of fitting scene lights on the sides and rear of the roof.

This 1970 Leyland Beaver/Carmichael fitted with Swedish-made Orbitor 72-foot hydraulic platform never made it into the paint shops. Its original red livery was carried over when the brigade and its appliances became part of Merseyside Fire Brigade in 1974.

This one was getting on in years when photographed in the early 1970s. Delivered in 1955, it is a Denis F15 series water tender based on a similar chassis to the popular F12 dual-purpose appliance but with a 400-gallon water tank and rear-mounted pump.

In 1965 Stockport received two Commer VA-based appliances, one an emergency tender and the other this smart water tender. Built to a John Morris & Sons design, it was relatively unique in design with only Buckinghamshire Fire Brigade operating similar types.

Another AEC Mercury turntable ladder but with a difference. This one dating from 1963 is one of the rarer types, fitted with a 100-foot Magirus ladder supplied by UK agents John Morris & Sons. Glasgow and Walsall operated similar appliances.

In the last years of Stockport Fire Brigade's independence, it commissioned five ERF appliances. This 1973 'Firefighter' model was one of two Stockport water tender ladders bodied by ERF's own Sandbach-based fire engine body-building division.

Walsall, formerly in Staffordshire, had two fire stations with a post-war fleet comprising almost entirely of Dennis-based appliances. Their turntable ladder was one of the exceptions. Delivered in 1962, this AEC Mercury was fitted with a 100-foot Magirus ladder and with the remainder of the fleet was incorporated into the West Midland Fire Brigade fleet in 1974.

Warley was another with a short-lived brigade. Formed in 1967 from the separation of Smethwick and West Bromwich fire brigades, it ceased to exist in 1974 when incorporated into the newly formed county of West Midlands. This Dennis F106 delivered in 1968 was one of the brigade's first new appliances.

Resplendent in the sunshine, Warley received this Perkins diesel-engined Dennis F108 pump escape in 1969. Wheeled escape ladders were becoming increasingly obsolete in this era and this one is seen carrying a Merryweather 50-foot steel version.

A relatively rare appliance and chassis combination here is Warley's 100-foot Carmichael-Magirus turntable ladder mounted on a Dennis F37 chassis. Similar models could be seen in Bury, Oldham and Surrey.

In the first decade after the Second World War, Warwick County Fire Brigade's fleet consisted of Rootes Group-, Commer-, and Karrier-based vehicles until 1962 when they commissioned this Dennis F26 water tender to cover the Leamington Spa district. It was the sole example in the Warwick fleet.

After the plethora of Commer and Karrier appliances those from Vauxhall Motors of Luton became the mainstay of the fleet. This Bedford TK water tender ladder with HCB-Angus coachwork was one of four delivered in 1971.

From 1969 Warwickshire's fire appliances were allocated registration plates with '999' digits. Still loyal to Vauxhall Motors, this 1971 Bedford TK was one of six 4.2-litre Jaguar-powered water tender ladders delivered in 1972.

A superb example of Britain's engineering craftmanship with the brigade title mounted on a varnished wooden board on the side of the ladder. It features a Magirus 100-foot turntable ladder mounted on a Commer chassis. The whole unit was later exported to Australia.

The Warwickshire fleet comprised of three ERF appliances with Simon Engineering hydraulic platforms. The first one, shown here, had 85-foot booms, while the others were slightly shorter at 70 feet. Delivered in 1968, this one was stationed at Leamington Spa.

HCB-Angus built several of this pattern of emergency tenders for British customers with Warwick County receiving this one in 1965 for Leamington Spa fire station. The chequered band identifies it as having the dual facility of a fire ground control point.

The Western Area of Scotland Fire Brigade governed six full-time and forty-five part-time and volunteer manned stations. Loyal to the early post-war Dennis models, there were five of these small Rolls-Royce-powered F8 water tenders in the brigade. This one, delivered in 1957, was based at Paisley.

A later model Dennis, this F26 series water tender dating from 1962 was one of a mixed batch of nine pump escapes and water tenders. The two wheeled carriage on the roof signifies that the appliance carried a Coventry-Climax portable pump and there is just a hint of high-visibility warning colours on the front bumper.

This 1968 Dennis F108 water tender devoid of any unpainted bodywork was one of the last three Dennis examples ordered by the brigade before it opted for its pumping appliances to be built on Dodge chassis. All of these Western appliances are still equipped with traditional wooden hook ladders.

Wigan Fire Brigade was one that continued the tradition of naming their appliances, in this case after members of the council or fire brigade committee. This 1965 Bedford TK/Hampshire Car Bodies water tender, named *Riley*, is as immaculate as the day it left the factory.

This 1970 Dennis F46 water tender ladder was the last appliance ordered by the borough before its fire fleet was incorporated into Greater Manchester in 1974. Identical to a series of appliances for West Riding of Yorkshire Fire Brigade, this one was named *Hanley*.

New red livery, painted tyres, traditional wooden ladder and chrome bells. Classic! Wigan's refurbishment of a pensioned-off 1954 Auxiliary Fire Service green goddess gave the appliance a new lease of life, at least until the brigade was incorporated into Greater Manchester in 1974.

As with Rochdale, Wigan was another of the Lancashire brigades that opted for German manufactured Metz ladders receiving this Dennis F21-based appliance with 100-foot ladder in 1957. This one displays the nameplate *Somers*.

Another of Hampshire Car Bodies emergency tenders, named *Ald Ball* after one of the council's aldermen, was delivered in 1962. It features roof-mounted strip lights for illuminating the surrounding area, a characteristic of many fire brigades in the Lancashire region.

The introduction of pedestrianised shopping complexes from the 1960s resulted in a relatively short-lived production of specialised appliances able to gain access to areas restricted to full-sized fire appliances. This low-height Branbridge-bodied Land Rover from 1969 was one of a pair in Wolverhampton.

The City of Wolverhampton was another that at one time operated a joint fire and ambulance service. For primary fire-fighting and rescue purposes the city commissioned this smart Dennis F108 appliance with 50-foot Simon hydraulic platform.

Alfred Miles & Sons of Gloucester built two appliances on Ford's Thames Trader chassis including this unique emergency tender supplied to Wolverhampton in 1962 shortly before Miles was absorbed into the Dennis Group. This appliance still survives having spent its later years as a mobile home.

Also available from Amberley Publishing

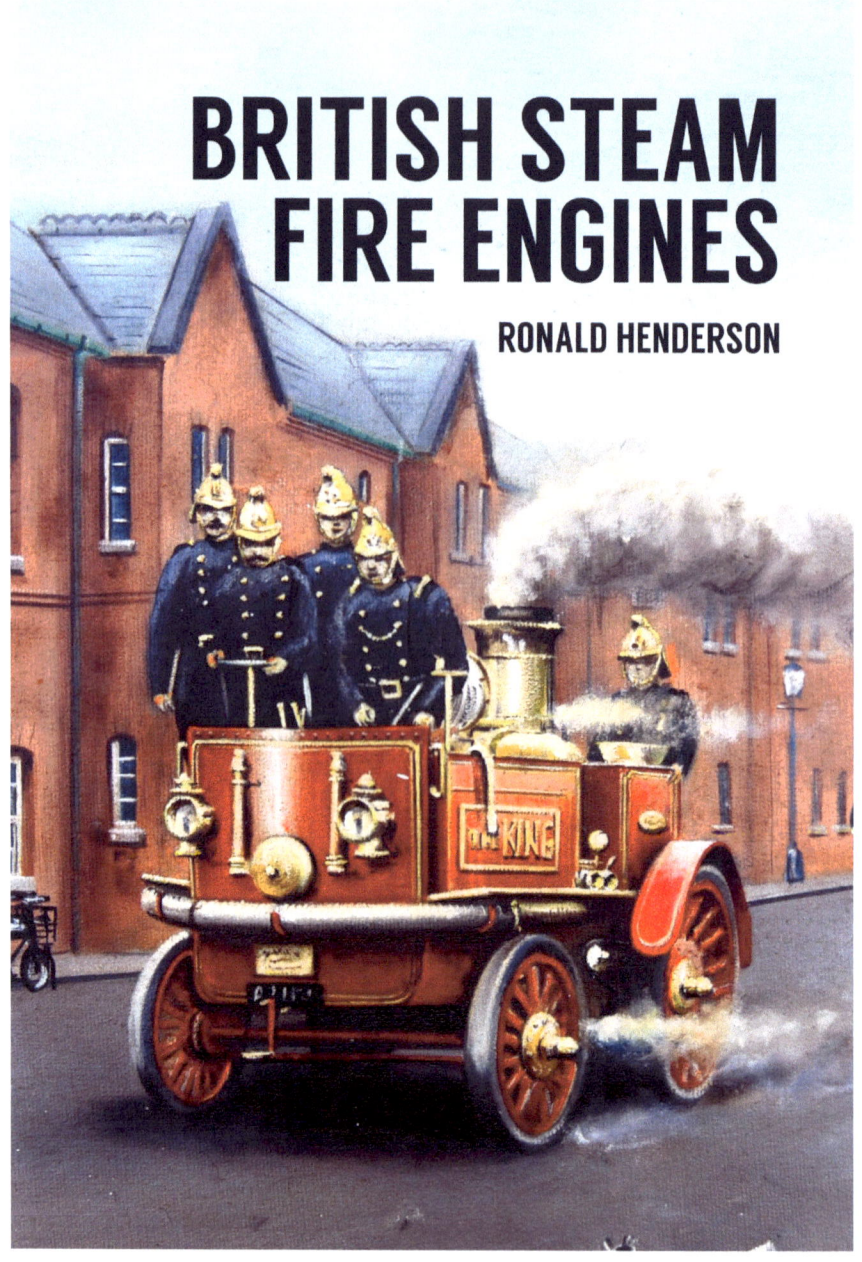

The fascinating story of the early steam fire engines.
978 1 4456 5779 0
Available to order direct 01453 847 800
www.amberley-books.com